Elementary Physics

Magnetism

THOMSON
GALE

San Diego • Detroit • New York • San Francisco • Cleveland • New Haven, Conn. • Waterville, Maine • London • Munich

THOMSON
GALE

For more information, contact
The Gale Group, Inc.
27500 Drake Rd.
Farmington Hills, MI 48331-3535
Or you can visit our Internet site at http://www.gale.com

Photo Credits: **The Brown Reference Group plc:** 2t, 3t, 3c, 4t, 4c, 4bl, 6; **Canadian Tourism Commission:** 1, 20; **Corbis:** James Leynse 18; **Leslie Garland Picture Library:** 16; **Hemera Photo Objects:** 2b, 4cbl, 4cbr, 4br, 13; **NASA:** 10, 14; **Photodisc:** Ryan McVay 12.

Consultant: Don Franceschetti, Ph.D., Distinguished Service Professor, Departments of Physics and Chemistry, The University of Memphis, Memphis, Tennessee

For The Brown Reference Group plc
Text: Ben Morgan
Project Editor: Tim Harris
Picture Researcher: Helen Simm
Illustrations: Darren Awuah and Mark Walker
Designer: Alison Gardner
Design Manager: Jeni Child
Managing Editor: Bridget Giles
Production Director: Alastair Gourlay
Children's Publisher: Anne O'Daly
Editorial Director: Lindsey Lowe

LIBRARY OF CONGRESS CATALOGING-IN-PUBLICATION DATA

Morgan, Ben.
Magnetism / by Ben Morgan.
p. cm. — (Elementary physics)
Includes bibliographical references and index.
ISBN 1-41030-080-3 (hardback: alk. paper) — ISBN 1-41030-198-2 (paperback: alk. paper)
1. Magnetism—Juvenile literature. 2. Electromagnets—Juvenile literature. [1. Magnetism. 2. Electromagnets.] I. Title.

QC753.7.M67 2003
538—dc21 2003002546

Printed and bound in Singapore
10 9 8 7 6 5 4 3 2 1

Contents

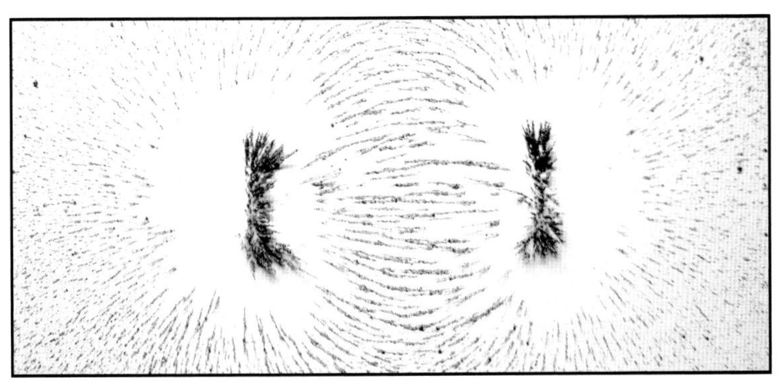

magnet and paper clips

A magnet attracts some steel paper clips (left). Which other items on this page do you think the magnet would attach to?

china piggy bank

pencil sharpener

rubber duck

wooden chess piece

metal bol

What Is Magnetism?

Magnetism is an invisible **force** that can push and pull objects without even touching them. Magnetism works through air and water. It also works through **solid** objects like walls and table tops. If you hold a **magnet** under a table, you can make things slide around on top, as if by magic.

Not all things are attracted to magnets. Try to pick up coins, marbles, plastic toys, and paper clips with a magnet. You will find that some stick, but others do not. Almost all the objects that stick to magnets are made of certain **metals**. Things made of steel, such as paper clips, are very magnetic.

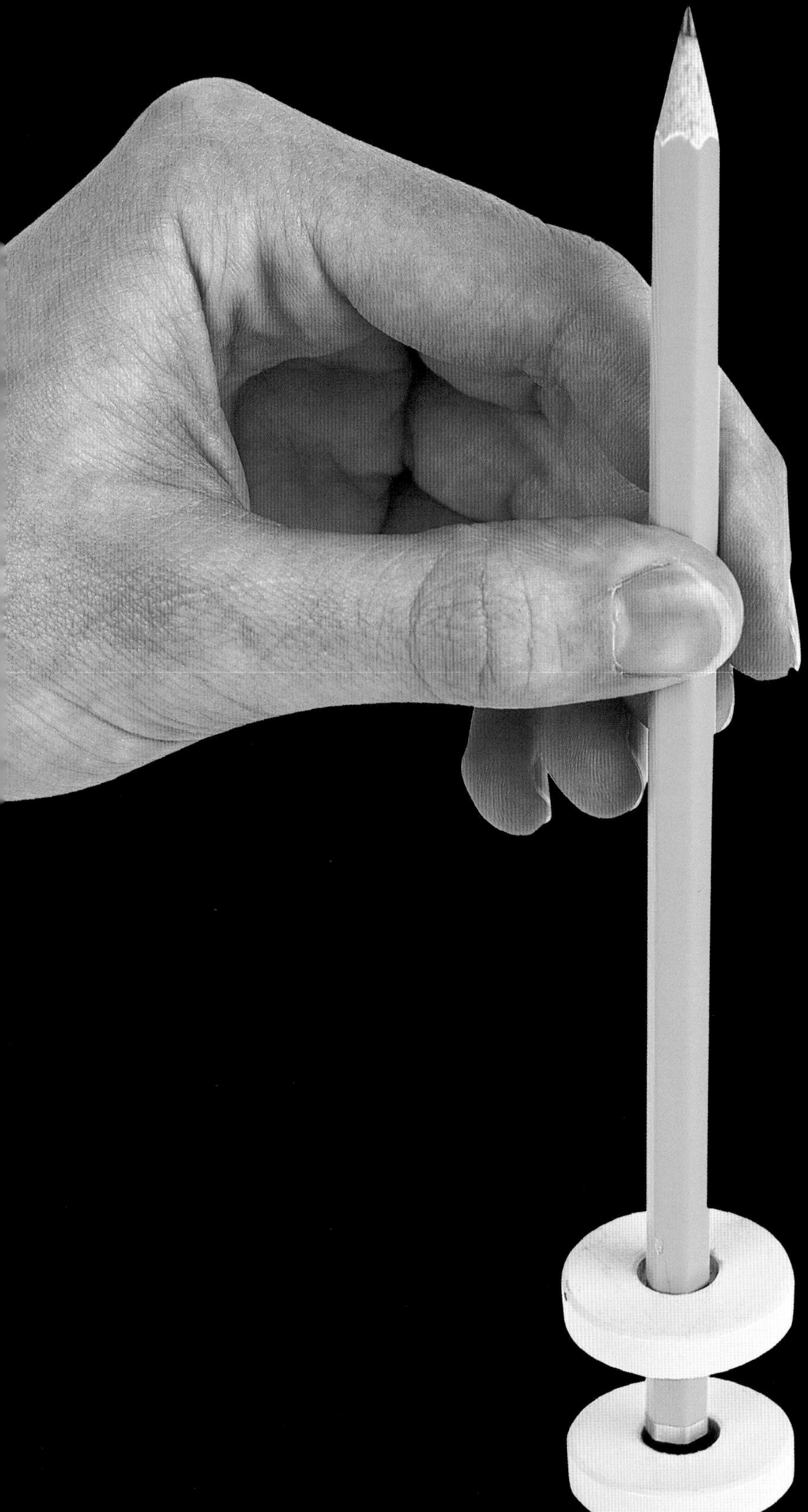

Place two ring magnets on a pencil with the same poles facing each other. The top magnet will float above the other one.

Poles Apart

Try pushing two **magnets** together. Sometimes they pull each other so strongly that one jumps onto the other. At other times, they push each other away. If you are careful, you can feel this pushing force in your fingers.

The two ends of a magnet are different. One end is called the north **pole**. The other end is called the south pole. Two north poles always push each other away. So do two south poles. Opposite poles attract each other, though. North always pulls south, and south always pulls north.

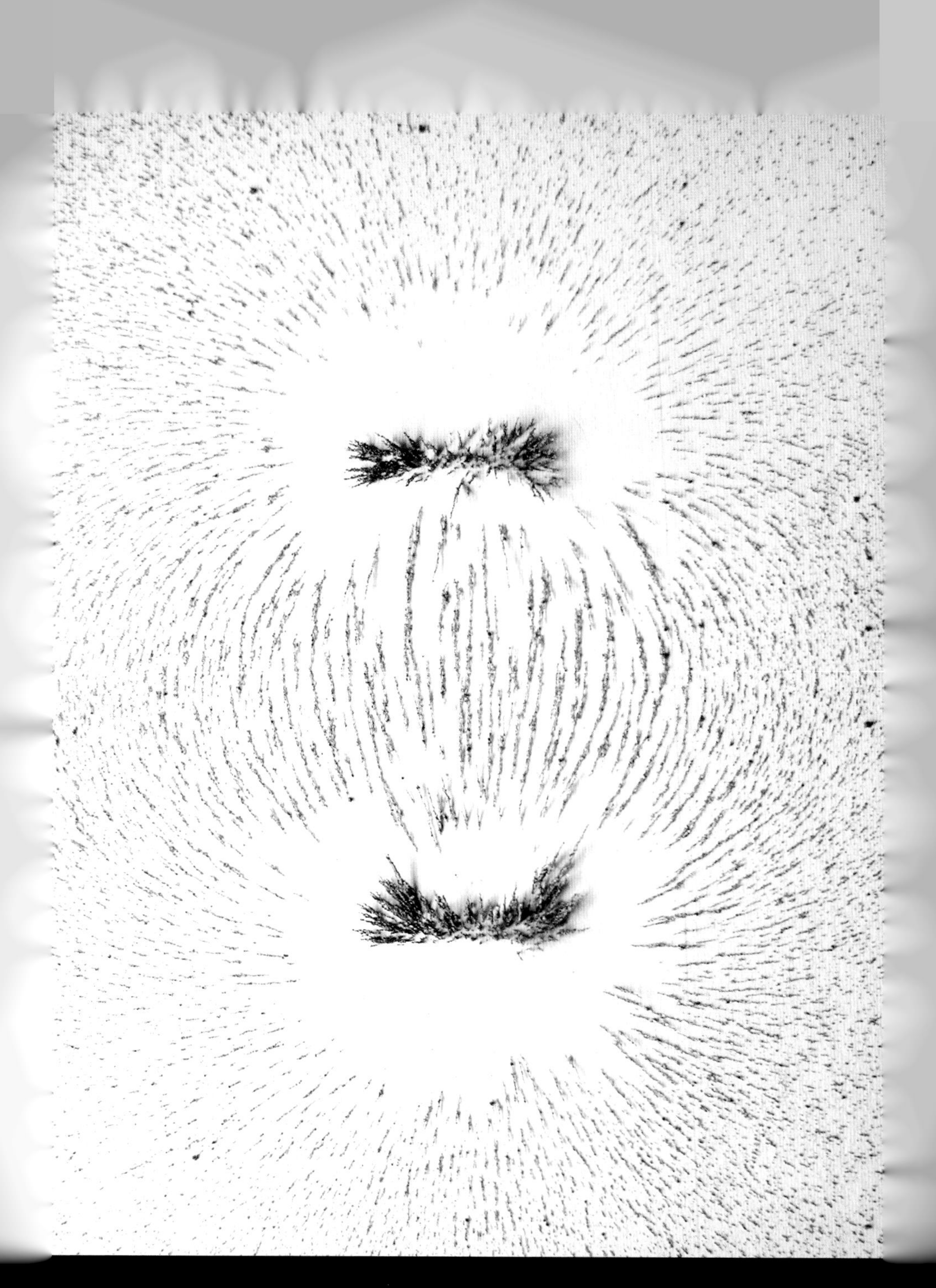

The magnetic field made by two magnets under a sheet of paper. The pattern of the magnetic field is shown by tiny flakes of metal called iron filings.

Magnetic Field

The **force** around a **magnet** forms an invisible pattern. This pattern is called the **magnetic field**. You can see the magnetic field if you have a magnet and some **iron filings**. Scatter a very thin layer of iron filings on a piece of paper. Then hold the paper above a magnet, without touching the magnet. Tap the paper gently. The iron filings will form a pattern of curved lines. These lines let you see where the magnetic field is. They show the direction of the pulling or pushing force that leads to each **pole**.

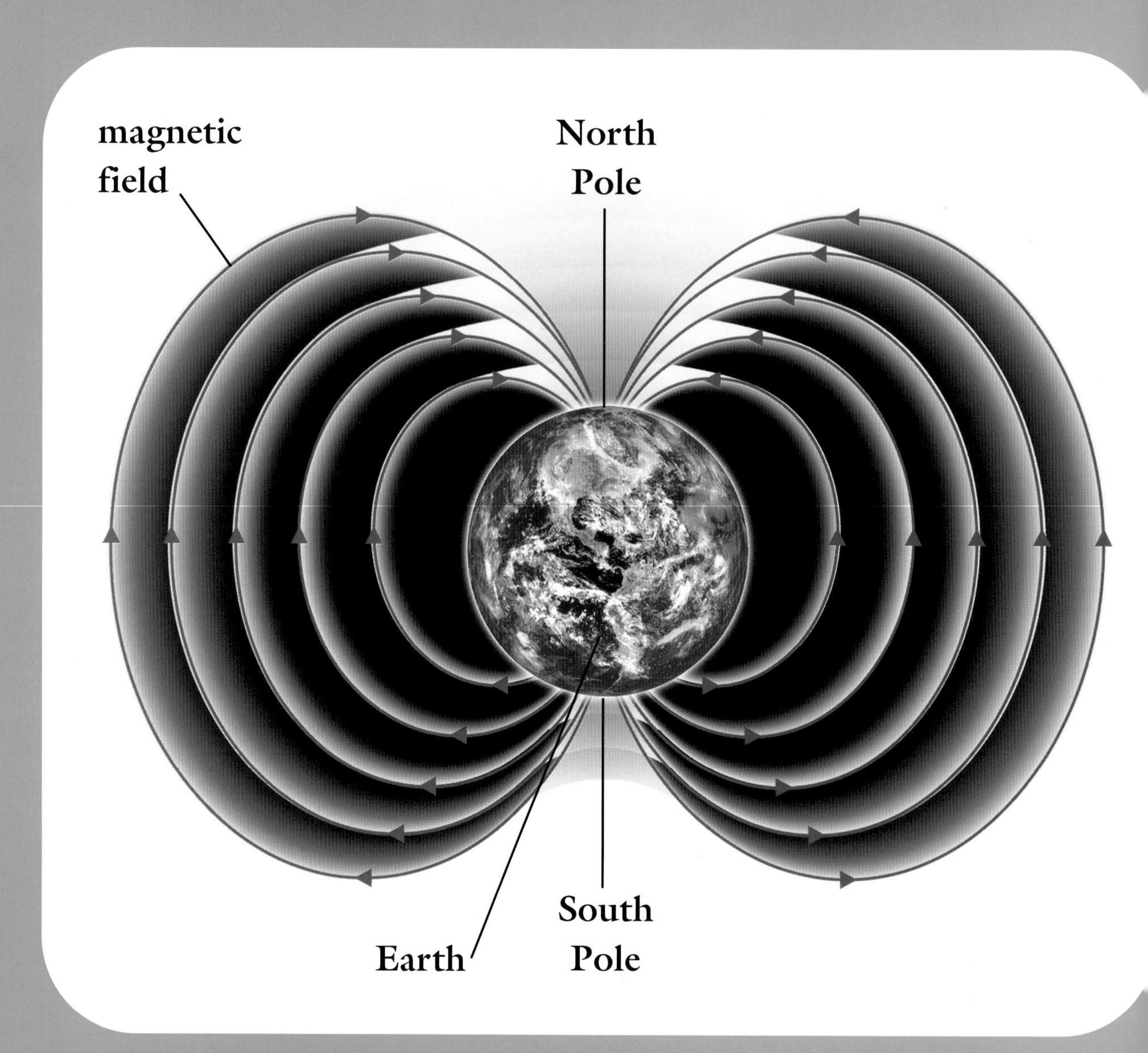

Earth is like a huge magnet. Its enormous magnetic field runs from its magnetic South Pole to its magnetic North Pole.

Planet Magnet

Earth is a gigantic **magnet** that is surrounded by a **magnetic field**. You cannot see Earth's magnetic field. But you can see its effect with a **compass**. A compass contains a tiny magnet that can swing around. One end always points to Earth's North **Pole**, while the other points to the South Pole.

You can make a compass with a **bar magnet**. Place the bar magnet on a styrofoam tray or something else that floats. Then float the tray in a large bowl of water. Keep the bowl away from metal objects. The tray will quickly swing around until the magnet points from north to south.

Turtles can sense Earth's magnetic field. This sense helps guide them on their travels through the oceans.

Finding the Way

Some turtles migrate, or travel, through Earth's vast oceans. They might have to swim hundreds of miles through the open water, with no landmarks to guide them. Yet somehow, they find their way.

honeybee

Scientists think that turtles and many other animals can do this because they sense Earth's **magnetic field**. This ability is like using a **compass**, so scientists say these animals have an internal compass. Salmon, honeybees, and dolphins all have internal compasses. So do birds that fly a long way south in fall.

Sun storms have a looped shape. That is because the exploding material flows along curved lines that make up the Sun's magnetic field.

The Sun's Magnetism

The Sun is a much bigger and more powerful **magnet** than Earth. Its **magnetic field** is more complicated, though. The Sun is a ball of burning **gas**, so its insides are always stirring around. This makes its magnetic field keep changing shape and getting tangled up. When the field gets very twisted, huge storms burst out of the Sun's surface and explode into space. These are called sun storms.

Every 11 years or so, the Sun's magnetic field gets so twisted that it falls apart. A new magnetic field then begins to form.

In junkyards, huge electromagnets are used to pick up cars.

Electromagnets

When **electricity** flows through a wire, a **magnetic field** forms around the wire. Electric wires usually make very weak **magnets**. But if the wire is wrapped around an iron bar again and again, it forms a strong magnet. This is called an **electromagnet**.

Electromagnets are very useful, because they can be switched on and off. Small electromagnets are used in all sorts of electric devices. Doorbells, hair dryers, washing machines, and battery-driven toy cars all have electromagnets.

A maglev train speeds over the track.
This train's powerful magnets make
it hover above the rail.

Floating Trains

One day, people might ride in trains that float. **Maglev trains** use powerful **electromagnets** to make them hover above the train rails. There are electromagnets on the bottom of the train and more electromagnets on the track. The **magnets** are arranged so that the same **poles** face each other. The poles push each other away, and this makes the train float.

Maglev trains can go much faster than other trains. They do not rub against the track, so they can shoot along at more than 300 miles per hour. Scientists are still working to improve maglev trains.

The northern lights glow in
the night sky over Canada.

Northern Lights

If you ever go to Alaska or northern Canada, look at the sky on a cold, clear night. You might see beautiful green and pink waves shimmering across the sky. These patterns are called the **northern lights**.

The northern lights happen because of Earth's **magnetic field**. Space is full of tiny particles that are attracted by Earth's **magnetism**. The particles crash into the sky around the North and South Poles. As they hit the air, they make it glow. Similar patterns that sometimes show in the sky around the South Pole are called the **southern lights**.

Magic Paper Clip

This experiment needs a steady hand! You need a paper clip, a length of cotton thread, and a magnet. Tie one end of the thread to the paper clip. Stick the other end to a table with a piece of tape. Carefully use the magnet to make the paper clip rise from the table, but do not let the two touch. Try to lift the paper clip directly above the tape and then down again.

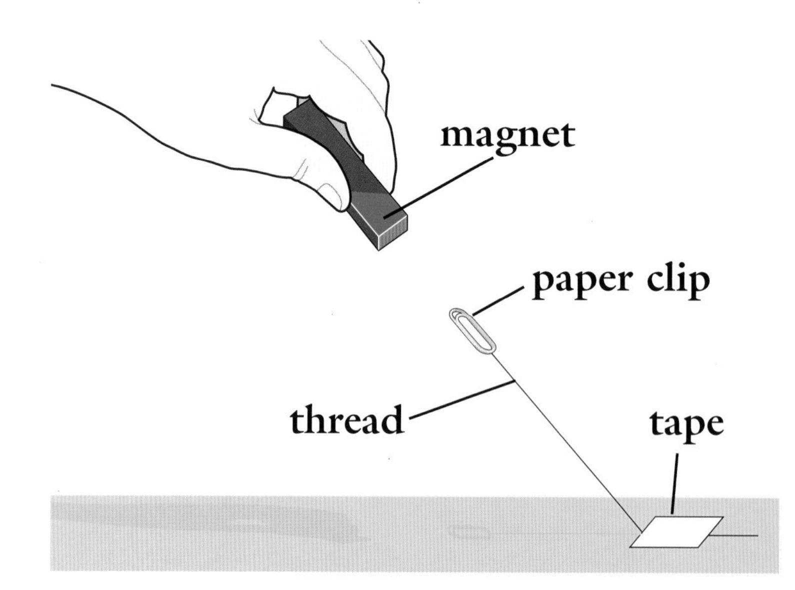

Glossary

bar magnet a long, thin magnet.

compass an object that points to Earth's magnetic poles.

electromagnet an object that is made magnetic by electricity.

force any action that changes the shape or movement of an object.

gas a substance that will spread to fill any space that contains it.

iron filings very small pieces of iron metal.

liquid a substance that can be poured. It will take the shape of the container that it is in.

maglev train a train that floats above the rail on a magnetic field.

magnet an object that can attract certain metals.

magnetism the attractive power that some objects have.

magnetic field the force around a magnet.

metal a hard, shiny, usually solid material, such as iron or gold.

northern lights bright patterns that sometimes show in the night sky over Canada and Alaska.

pole one end of a magnet. The North Pole and South Pole are at the ends of Earth's magnetic field.

solid a hard substance that keeps its shape.

Look Further

To find out more experiments you can carry out with magnetism, read *101 Great Science Experiments* by Neil Ardley (DK Publishing).

You can also find out more about magnetism at this website: www.howstuffworks.com/compass.htm

Index